GOD'S HOLY BOOK OF NUMBERS

Pam Atwater

Sources of scriptures are referenced from:
The ESV Bible App. Crossway Books, Version 5.0.6., 2010.
www.esv.org

Copyright © 2024 Pam Atwater
All Rights Reserved.
HerLife HerWrite Publishing CO. LLC
ISBN: 979-8-9889954-2-5

Introduction

My book is inspired by the Holy Spirit. All information has been given by the Holy Spirit. These numbers are not "angel" numbers. Those that have an ear let them hear what the spirit says to the churches. In my book of numbers, it gives you a glimpse as to some things that the Lord is speaking to you through these numbers. May the Lord Jesus Christ continue to give you more insight and revelation as he speaks to you through these prophetic numbers.

Table of Contents

Number 1	7
Number 2	10
Number 3	14
Number 4	18
Number 5	21
Number 6	24
Number 7	27
Number 8	30
Number 9	32
Number 10	35

Number 1

Any number that is speaking to you or is standing out to you, you must have the word of the Lord inside of you for you to understand what the Lord is saying to you concerning that number.

When the Holy Spirit is speaking to you concerning the number 1 the Lord is directing you to a call of faith. The number 1 also represents the first day that the Lord created the light. It also means that you are not to have another God before the Lord and Savior Jesus Christ. Genesis 1 verse 11 stands for New Beginning because everything starts with the number 1. The number 1 is also a whole number which stands for wholeness and the Lord wants us whole in him, complete in him because Jesus is the only one that can fill us and make us whole. The number 1 also means the trinity, which is The Father, Son, Holy Spirit because they are one. The number 1 also represents in the bible that 2 becomes 1 in marriage.

Philippians 1 verse 11 speaks of being filled with the fruit of righteousness, spiritual wisdom, blessing, manifestation, and alignment because the 11 is lined upward, straight up, and down. Genesis 1 verse 11 also talks about seeds. The Lord wants us to practice the fruit of the spirit so we can produce good fruit. Also, In the book of Romans Ch 1 verse 11 it speaks of how the Lord reveals in you the spiritual gifts and spiritual wisdom, spiritual sight so you can see into the supernatural. The Lord is going to show you what is already inside of you with your gifts, spiritual gifts, whether it is prophesying, teaching, even ministering or singing etc. The Lord Jesus is manifesting greater things in your life, greater things in your life than what you had to leave in the past. The Lord is about to open doors for you, doors of opportunity, success, and favor according to Ephesians 1 verse 11.

    In the book of Luke Ch 11 verse 11 it is stating that the Lord is only going to give you his best once you press into his presence. Deuteronomy 11 verse 11 talks of rain from heaven. It is simply saying blessing, showers of blessing. Even when it seems like you have been waiting for a while, when you start to see these numbers, that means that the Lord is about to blow your mind. The Lord wants you to have faith in him and to trust him that he is a rewarder of them who seek him.

~Number Notes~

Number 2

The number 2 stands for the miracles and wonders of almighty God. Even as we begin to understand the number 2 or 222 the numbers will mean something to you right now and then later in life, they will mean something else for you as you see these numbers standing out to you. One of the first things that the number 2 stands for is being a faithful witness for the Lord Jesus. As ambassadors for the Lord, he requires us to be his hands and feet on the earth, proclaiming his name doing the work of the Lord which is witnessing to others. The number 2 also means you must be separated from the world; you cannot act like the world. There is a difference between righteous and unrighteous.

Number 2 also represents the Lord's open doors for you and the authority to decree the Lord's purpose, doors of favor, abundance and for you to take new grounds of territory. When the Lord is presenting you with open doors and opportunities the Lord is simply saying to you that you can trust the Lord with open doors, as well as closed doors.

In the book of Isaiah Ch 22 verse 22 the scriptures that the Lord is speaking about is authority. As believers we have the authority to bind and to lose whatever the enemy tries to throw at us. Matthew Ch 16 verse 19 says, "I will give you the keys of the kingdom of heaven and whatever you bind on earth shall be bound in heaven and whatever you lose on earth shall be loosed in heaven." This is the scripture concerning the keys of our father's house, the kingdom. We are the Lords Sons and Daughters and if we do not have the keys, how will we know our inheritance? We need to know what those keys are to unlock doors, shift, and move things from heaven into the earth.

Also, in Isaiah Ch 22, when it talks about our shoulder, the enemy wants to place on God's people burden, shame, and a heaviness of this world that is why the Lord says that he is placing upon our shoulders the keys of David. So, when the Lord speaks of open doors those open doors only come by being at the feet of Jesus as well as obedient. When we are not in our secret place with the Lord you lose your keys, because when we are so focused on the cares of this world, and the heavy weight that the enemy wants us to focus and we take our focus off the Lord, we lose our keys.

In Hebrew, the number 2 or 22 means completion. It means that you are about to step into what you have been stored in the secret place with the Lord. The number 2 also means that there are a lot of promises that the Lord is about to give you. In the book of Genesis Ch 22 verse 2 it speaks of the word love in the bible. The number 2 speaks of how 2 shall become 1.

~Number Notes~

Number 3

When it comes down to the number 3 or 333 immediately the Holy Spirit is telling us about the Father, Son, and Holy Spirit. In Hebrew, the number 3 means Shelosh and Shelosh means Harmony, New Life and completeness. The number 3 is such a great number. The number 3 could very well mean that you are getting goals in your life to be very prosperous, and the creative process are all coming together.

When it comes to the number 3, the opposite of 3, would be that there is a lot of chaos going on and that you are trying to get things in order, but the enemy is fighting you on every side, but for the most part, the number 3 is a very fruitful and positive number. Jeremiah verse 3 says, "call to me and I will answer you and tell you great and unsearchable things you do not know. That means the Lord is trying to get your attention. So, there could be a completion to a thing that the Lord is about to take care of.

The number 3 also stands for the way, the truth, and the life. The Lord Jesus always had his disciples with him, but it was only 3 disciples that he was close and that was Peter, James, and John. The number 3 also stands for a 3-day fast like the one Esther went on. When it comes to the number 3 or 333 that means it would be telling you that you are trying to get the Lord's attention because it is something that you do not know like, a sense of urgency. Like a 911 emergency, for example it could be something that you knew that could change your life for the good. Ezekiel 33 verse 3 says, "And he says a sword coming against the land and blows the trumpet to warn the people." So, it could be letting you know that the lord needs to talk to you about something, very urgent information to even save your life.

    The Gematria for the number 3 means help, heal, and be held. Number three also means that you are moving forward, whether it is from a relationship, or a job, and you are doing well on this journey of a spiritual awakening. The number 3 also means that you are feeling like you are losing your natural abilities to be creative, but in fact the Lord Jesus Christ wants you to use whatever the Lord has given you wisdom to use, which is your natural abilities of what the Lord has given you to use that is inside of you. It could also be that the Lord wants you to use when you are in a storm, he wants you to use, use that to help you bring that fire and spark back into your creativity.

Also the number 3 could be getting you to be aware of more of your gifts are being activated, and that you are growing out of the old you and coming into the new you, where the Lord wants to show you more, whether it's writing a book, working on your health, ministry of music, even family etc. the Lord is giving you the natural ability to do it. And the Lord wants you to know that you are not alone. The number 3 means changes are coming and for you not to allow fear to come up on you because fear will try to come into your mind, but you are to blot out those negative thoughts with the word of the Lord by reading it daily.

~Number Notes~

Number 4

The number 4 automatically means a separation. Put a separation from the sheep and the goat, Goshen from Egypt. The Lord is distinguishing his children from the ones that are not. In Matthew chapter 25 the Lord was doing a lot of separation from the faithful ones to the ones that were not faithful separation from the people that says they are born again, and from the ones that are actually living right and are born again this would be a time where the lord is doing heavy examining, being tested in the hearts on their performing. The number 4 also means being set apart for God's purpose and made in the Lord's image.

In the book of Numbers Ch 4 verse 44 and Luke chapter 4 verse 44 is all about being at a higher power, set apart. John chapter 4 verse 44 is a prophet being set apart in his own hometown because a prophet has no owner in his own country. The number 4 means the preaching of the gospel. It also means creative works, the number of creations that the Lord completed making the universe on the fourth day. The Lord brought into existence the sun, moon, and the stars. Their purpose was not only to give off light, but to divide the day from the night which becomes seasons and time. Daniel chapter 2 verse 20 through 21.

The number 444 could also mean you are experiencing big changes in your life which the Lord could be guiding you in a new direction. The number 4 also means honesty and justice and justice will be served. It also means that you see your gifts that are inside of you and for you to recognize people and your true character. You are rarely fooled by people, and you have the gifts to read body language or even hear what people are thinking. The Lord is also letting you know that you have an abundance of love in your life whether you know it or not.

~Number Notes~

Number 5

The number 5 or 55 means grace. The number 5 also means the 5-fold ministry. The number 5 means your 5 senses. 5 also means the ability to overcome something in your life that is put as a stumbling block. Jesus receives 5 different wounds on the cross. His head, both hands, both feet, and his sides. Now we are overcome by the blood of the lamb and by the words of our testimony. In the bible Israel came out of Egypt in ranks of 5 and it was proven that it was God's power that gets all of them free. So, for me and you, in the bible the number 5 or 55 means you could be headed into your promised land and there's uncertainty yet excitement.

When you see the number 5 it means that the Lord is and will be smoothing things over and it is all going to be all right. On the fifth day in the bible the Lord created the birds and the fish. When it comes down to 5 or 55 of any other number sometimes that number will mean one thing to you in one season of your life and another thing to you in another different season of your life. The number 5 again is the Lord's grace that he is given you the ability to overcome everything and anything that has come into your life. Understand that if the Lord brought you to it, the Lord could see you through that and any situation that you are being faced with.

The Lord has given you the grace to fulfill it and to get it done well. It is the Lord's favorite on you, Isaiah chapter 55 verse 5 says, "surely you will summon nations you know not, and nations you do know will come running to you, because of the Lord your God, the holy one of Isaiah, for he has endowed you with splendor." Understand when the grace and the favor and the anointing is upon you, it is not because of your good works, or what you have done or how good we believe. It is because of the Lord your God and his grace, the move of God, the hand of the Lord and the plan of the Lord.

~Number Notes~

Number 6

When it comes down to the number 6 or 66, it means opposition against the Lord in what man is choosing to do that the Lord Jesus Christ does not approve of. It means that man is always in opposition with the flesh and spirit. Galatians chapter 5 verses 16-18 says, "this I say then, walk in the spirit, and ye shall not fulfill the lust of the flesh." Verse 18 says, "But if ye be led of the spirit ye are not under the law." And when it comes to our flesh the spirit of Jesus Christ is the one for you to seek for all your answers. Your flesh is an enemy to Gods plan and for the Lord's purpose on your life.

The number 6 also mean for you that you need to align your thoughts with what the mind of Christ has for you. Sometimes you have to ask yourself are you going through transition or are you going through some type of change. Sometimes our body and our mind go through things that we do not quite understand, and again that is when we need to make sure our mind, our thoughts, and actions line up together. When it comes to the number 6 it is always best that you ask the Holy Spirit what does it means for you when you see the number 6 or 66. We must understand that in every season you are in your life that it is a part of your journey and in that season of your life there is a lesson and a purpose in your life that needs to be learned. Also, this part of your life must happen so you can get from one step then to the next.

~Number Notes~

Number 7

When seeing the number 7 or 777 it stands for completeness or wholeness. The number 7 also means that the Lord Jesus is doing things by his Spirit. It also stands for how the Lord created heaven and the earth and on the 7th day the Lord rested. When it comes down to the word rest, the Lord Jesus Christ wants us to do just that, meaning He the Lord wants to be our rest, our hope, our joy and our peace and not to worry about the cares of this world, because Jesus is our hope, so when you see the number 7 it means to look to Christ for our hope.

The number 7 also means purification, and perfection as well as forgiveness. In the book of Matthew chapter 18 verse 21 and 22 the number seven is also the seal of the Lord Jesus Christ. If you are constantly seeing this number, then the Holy Spirit is telling you that you have a blessing that the Lord wants to give you and that it is on the way. Number 7 is the foundation of the word of the Lord, his authority. 7 also represents that your heart's desires are coming to pass. It also means that you are on a journey of growing spiritually and new opportunities are unfolding, and you are in alignment and open to what the Lord wants to give to you. As it pertains to the number 7 of wholeness the Lord Jesus wants you to be whole within yourself not broken, but whole and resting in Him, wholeness in your heart, and wholeness that you know who you are and that you who the Lord has called you to be and not the world.

~Number Notes~

Number 8

When it comes down to the number 8 the number 8 stands for a new and renewed mind and a renewed heart, and a renewed spirit. The number 8 speaks of a new beginning, a new mindset and a new love for your Lord and savior Jesus Christ. When you see the number 8 or 888 it also speaks of the new commandments that the Lord has given us to follow in the book of John Chapter 13. The number 8 also stands for a cleansing that is taking place.

In the bible in the Old Testament there were only 8 people in the boat of the Ark. So, it was only 8 people in the boat because the Lord wanted to spare them because the Lord wanted to start something new. Also, during the resurrection of Christ, Christ appeared 8 times after he was resurrected. Genesis Chapter 8 also speaks when Noah and his family came out of the Ark. So, for us, the Holy Spirit is letting us know that we are coming out with a new beginning and a new season and a new lifestyle, also including if you are dreaming of the number 8 it means that you are about to give birth to a new business, a marriage or some type of new relationship etc.

~Number Notes~

Number 9

The number 9 speaks of bearing fruit. Number 9 speaks of manifesting the Holy Spirit and the Holy Spirit training process by learning the difference between right and wrong, good and evil in the believer's life, therefore while the Holy Spirit is teaching and exposing what is in the heart. There are many examples in the bible such as the nine gifts of the spirit which are the word of wisdom, the word of knowledge, faith, healing, miracle, prophecy, discerning of spirits, tongues, and the interpretation of tongues, which can be found in the book of 1st Corinthians Ch 12 verse 8 - 10. In the Holy Bible there are also the fruit of the spirit which can be found in the book of Galatians chapter 5 verse 22-23, and they are love, joy, peace, patience, kindness, goodness, faithfulness, gentleness, self-control.

There also is the beatitude in the book of Matthew Ch 5 verse 3-11 which are Blessed are the poor in spirit for theirs are the kingdom of heaven, blessed are those who mourn for they will be comforted, blessed are the meek for they will inherit the earth, blessed are those who hunger and thirst for righteousness, for they will be filled. Blessed are the merciful for they will be shown mercy. Blessed are the poor in heart for they will see God. Blessed are the peacemakers, for they will be called children of God. Blessed are those who are persecuted because of righteousness, for theirs is the kingdom of heaven, and blessed are you when people insult you, persecute you and falsely say all kinds of evil against you because of me. The number 9 also stands for being in an agreement with the Lord. The number 9 also means Judgment when we do not align ourselves with the Lord and the Holy Spirit.

~Number Notes~

Number 10

The number 10 means the perfect order of God's plan and purpose. It means God is setting his plan according to his will. The Lord is revealing his plan according to what he has been trying to show you or he is showing you. For example, when it comes to the 10 commandments, which is his perfect will being set up in perfect order and in the completion of a season and in that season things will start to manifest. John Ch 10 verse 10 says, "The thief only comes to steal and kill and destroy I have come that you have life and have it to the full." Which means, where the enemy has come, brought chaos, and destruction, the Lord almighty is speaking life of abundance to the believer and bringing into his perfect order so you can live a victorious and abundant life.

The number 10 also means the 10 plagues because in the bible when Pharaoh refused to listen to Moses or the Lord, God almighty sent those plagues to remind Pharaoh that he is not God. The number 10 is also a number representing how Noah was the 10th descended from Adam. The 10 commandments are also found in Exodus, as well as there are 10 statements in the bible of the book of John stating the words "I am". When it comes to the number 10 putting the Lord Jesus Christ first represents the number 10. We as God's creation have 10 toes so to walk with the Lord must be blameless.

~Number Notes~

~Number Notes~

~Number Notes~

www.ingramcontent.com/pod-product-compliance
Lightning Source LLC
Chambersburg PA
CBHW071501160426
43195CB00013B/2177